£ 14. 99

C000078526

CUMBRIAN STEAM

Gordon Edgar

AMBERLEY

Front cover: Royal Scot Class 4-6-0, 46115 *Scots Guardsman*, and the Border City of Carlisle go together like a hand in a glove, as will be revealed within the pages of this book. The revered locomotive looks as though it means business for the 47 miles of predominant climb ahead of it to Ais Gill summit as it approaches Petteril Bridge Junction with the return 'Waverley' charter to York on 22 July 2012.

Back cover: Royal Scot Class 4-6-0, 46115 *Scots Guardsman*, hurries away from the Border City, crossing the Eden River bridge at Etterby on 10 May 2011, working a morning positioning move from Carnforth to Cadder Yard.

First published 2014

Amberley Publishing
The Hill, Stroud
Gloucestershire, GL5 4EP

www.amberley-books.com

Copyright © Gordon Edgar, 2014

The right of Gordon Edgar be identified as the Author of this work has been asserted in accordance with the Copyrights, Designs and Patents Act 1988.

ISBN 978 1 4456 3962 8 (print)
ISBN 978 1 4456 3974 1 (ebook)

British Library Cataloguing in Publication Data.
A catalogue record for this book is available from the British Library.

Typeset in 9.5pt on 12pt Celeste.
Typesetting by Amberley Publishing.
Printed in the UK.

Introduction

The main lines snaking through Cumberland and Westmorland have for decades challenged generations of photographers, and the legendary locations of Scout Green, Shap Wells and Ais Gill have been witness to many eminent railway photographers over the years. The Border City of Carlisle continues to attract steam-hauled charter trains, indeed several times in a week during certain periods of the year. Even during the 'famine' years of steam immediately after the last fires had been dropped in August 1968, Carlisle was visited by the *Flying Scotsman*, the only steam locomotive permitted to run on BR metals at that time. Such was the popularity of steam traction that it was not to be long before regular charter trains were seen again on the Settle–Carlisle and Cumbrian Coast lines, and indeed Cumbria has never really been without steam traction on the main line for any great period of time. Who would have believed back then in 1968 that so much steam activity would have been possible in the twenty-first century, and not least on the West Coast Main Line over Shap summit? This book documents the contemporary scene, interspersed with a step back in time to the 1950s and 1960s, using historical monochrome images. These cover an incredible period of change on our railways and make for an interesting comparison with today's lines. I make no apology for including quite a number of archival photos taken on Kingmoor and Upperby depots, opportunities sadly no longer possible today, and feel that this provides a more balanced presentation. As well as being a tribute to the 'Men of Steam' who continue to work miracles on the main line with locomotives introduced into service no less than fifty years ago and in many cases much more, this book also portrays the steam locomotive in a contemporary environment with no attempt to recreate the past.

Carlisle has been, and continues to be, a magnet for the rail fan and excursion traveller, and it is only right that this important railway centre should receive adequate coverage. Main lines approached the Border City from almost every point of the compass, offering a huge variety of motive power, as the city route layout of 1963 demonstrates; this was a time before great rationalisation and shortly after the opening of the new marshalling yard at Kingmoor. My earliest memory of Citadel station is from the age of five, just before the original roof was cut back to its present overall coverage. Waiting with my parents on Platform 3 to return to London, I vividly recall the kaleidoscope of light coming through the overall roof and the amazing smoke-filled atmosphere beneath. It was this early experience that undoubtedly got me 'hooked' on railways, and as soon as I was able to travel alone I would return to Cumberland to stay with my grandmother for the summer school holidays. I hope that these archival and more recent photographs taken by myself convey to the reader some of that magic of the past and, against all odds, what we can happily still witness to this day, despite the changes on our railways. Long may the steam locomotive continue to provide pleasure to the many who travel behind it and flock to the lineside to witness its passage over the varied routes through the county to and from the Border City.

Gordon Edgar, Carlisle, July 2014

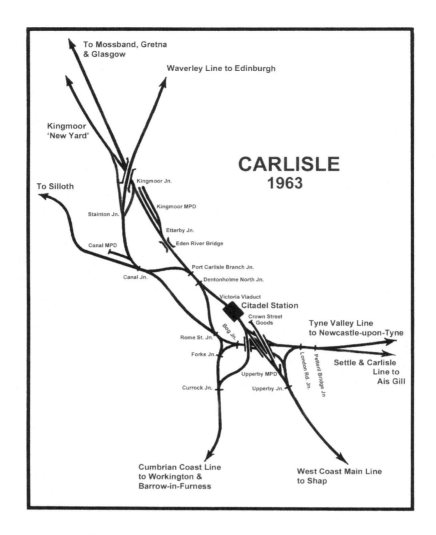

Above: Map of Carlisle's rail routes in 1963.

A very warm welcome to Cumberland! An early 1950s scene at Whitehaven Bransty station, with Ivatt 2MT 2-6-0 No. 46459 of Workington (12D) shed waiting for the road back to its home depot. Beyond the locomotive can be seen William Pit, which ceased winding in 1954.

The classic scene at Etterby Junction, Carlisle, which many enthusiasts will fondly remember, now transformed almost beyond recognition. On a winter's afternoon in the 1950s, Stanier Black 5 No. 45428 works south past Kingmoor depot, virtually obliterating it in its exhaust, in charge of an Up empty mineral working. Happily the mixed traffic 4-6-0 locomotive is still with us today on the North Yorkshire Moors Railway, named in honour of the eminent railway photographer the late Rt Revd Eric Treacy, Lord Bishop of Wakefield, 1968–76.

No. 46251 *City of Nottingham* stands at Platform 3 of Carlisle Citadel station heading 'The Duchess Commemorative' RCTS special on 5 October 1963, with the loco working throughout from Crewe to Edinburgh Princes Street and back. The charter participants form an orderly queue in anticipation of access to the footplate. The Princess Coronation Pacific was to remain for just one more year in service, being scrapped at Cashmore's, Great Bridge, during December 1964. (Author's Collection)

The classic view at the south end of Carlisle Citadel station with 46233 *Duchess of Sutherland* and West Coast Railway Company 47854, bearing its new name *Diamond Jubilee*, respectively heading the return 1Z79 16.01 Carlisle–Leicester 'Hadrian' charter and 'The Statesman' to Newport on the Queen's Diamond Jubilee weekend, Saturday 2 June 2012.

A classic pose of the 'Royal Scot' at Carlisle Citadel station, with 46233 *Duchess of Sutherland* sitting at the head of a full set of BR maroon Mark 1 coaches shortly after arriving with the 1Z31 06.19 Milton Keynes–Carlisle 'Royal Scot' charter on 9 June 2012. The 'Royal Scot' name was conferred in 1927, although a headboard was not carried by train locomotives until 1950. This was the first style of board, rectangular with semicircular ends and a Hunting Stewart tartan background.

With the sanders in full operation, the driver of Princess Coronation 4-6-2 46233 *Duchess of Sutherland* skilfully gets his train of ten trailing bogies underway on the wet rails upon departure from Carlisle with the 1Z33 16.13 Carlisle–Milton Keynes return 'Royal Scot' charter on 9 June 2012.

Gresley A4 Pacific 60009 *Union of South Africa* shunts the stock of the 1Z55 05.48 Worcester Foregate Street–Carlisle 'Cumbrian Mountain Express' (CME) at Victoria Estate Road, just to the north of Carlisle station, on Monday 27 May 2013.

No. 46226 *Duchess of Norfolk* heads an Anglo-Scottish express away from Carlisle. The twelfth-century red sandstone cathedral and Priory of St Mary can be seen through the locomotive's exhaust, while the stone building on the left of the photograph is the former L&NWR Grain Shed within Viaduct Goods. The site of the city's former gasworks was on the waste ground in the foreground. The Austin A40 Farina car would date this photograph as post-1958. (Author's Collection)

Above: With Dixon's chimney and mill and Carr's biscuit factory standing proud above the skyline, Jubilee 4-6-0 45699 *Galatea* raises the echoes as it lifts the empty stock of the 1Z86 07.08 London Euston–Carlisle winter 'CME' away from Carlisle station, at English Damside on 8 February 2014. The surviving BR-style concrete lamp posts on the now-disused extended part of bay Platform 8, just beneath Victoria Viaduct, are noteworthy.

Right: Looking down the cobbled Bush Brow onto English Damside on the lower level alongside the east side of Carlisle station, 45305 shuffles beneath Victoria Viaduct and through Carlisle station en route to Upperby for servicing, before returning the 'Fellsman' charter to Lancaster on 22 August 2012. Victoria Viaduct was formally opened by HRH Princess Louise on 20 September 1877. A former bank on the viaduct above that was converted into a jeweller's around 1959, and was reputedly haunted. Footsteps could be heard on the tall flight of steps within the building, but nobody could account for them. They finally stopped when the floor was lifted and an old well uncovered. The well had been partially filled in, and when they were emptying it a headless skeleton was discovered. It was removed and is now on display in the Tullie House Museum.

Seen from the Victoria Viaduct, Britannia Class 7MT Pacific 70052 *Firth of Tay*, with its cylinder drain cocks open, makes its departure from Carlisle Citadel heading a Perth service on 7 May 1966. The Pacific lasted just one more year in service, being withdrawn from Kingmoor in April 1967 and disposed of at Campbell's of Airdrie during November 1967. (Mike Blenkinsop, Author's Collection)

Obliterating the surroundings in a cloud of steam and smoke, 45699 *Galatea* draws the empty stock of the 1Z45 'Cumbrian Jubilee' stock from Tyseley Warwick Road away from Carlisle station on 9 November 2013, before proceeding to Upperby depot for servicing.

On 26 May 1967, a BR Horwich-built Black 5 4-6-0, 44674 of Kingmoor shed, stands in the extended Platform 8 beneath Victoria Viaduct at Carlisle Citadel awaiting departure with the 20.32 passenger service for Perth, by this time the last regular steam-hauled passenger steam working into Scotland. No. 44674 was withdrawn in December 1967, having seen just over seventeen years' service. (Author's Collection)

The Bush Brow, Backhouses Walk and English Damside area of Carlisle has always been a poor relation to the adjoining West Walls and the cathedral grounds, but the infrequently visited area at the foot of the city walls has for centuries been a convenient dumping ground. The urban detritus sets the foreground scene for Jubilee 45699 *Galatea* shunting empty stock across the arches in the rain on Saturday 8 February 2014.

Carlisle station retains a wealth of period features and, despite the modern-day signage, very little has changed over a period of more than half a century, including the jointed track and the 1958-installed canopies extending beyond the end of the cut-back overall roof. All that is required is a locomotive of Stanier design and the clock is turned back. An 8F 2-8-0, 48151, rolls into Platform 3 with the late-running 1Z70 09.12 York–Carlisle 'Waverley' on Sunday 25 August 2013, a bank holiday weekend, with the platforms virtually deserted.

A steam-hauled departure north from Carlisle is always a spectacle, the loco having to work hard to lift its stock away from the station, and the sound of the exhaust invariably reverberating loudly off the surrounding city-centre buildings. Here, 70000 *Britannia* is glimpsed from Victoria Viaduct Road as she draws the whole empty coaching stock formation of the 'Cumbrian Guardsman' charter out of Platform 1 prior to setting back into the coaching stock sidings adjacent to the west side station screen wall on 2 March 2013.

Almost conveying an everyday scene of the 1960s, Armstrong Whitworth-built 45305 rolls into Platform 3 at Carlisle in the heavy rain on 6 August 2012, heading the 1Z80 'Mersey Moorlander' charter, the 06.06 from Crewe and 07.19 from Liverpool Lime Street. Stanier Royal Scot 4-6-0, 46115 *Scots Guardsman*, declared a failure on the previous day, waits for a tow back to Carnforth depot for remedial attention.

Jubilee 4-6-0 45583 *Assam* brews up in the bay platforms at Carlisle Citadel on 29 July 1964. A Warrington-allocated locomotive, it was withdrawn from service just three months later, and disposed of at Central Wagon Co., Ince, during July 1965. (Author's Collection)

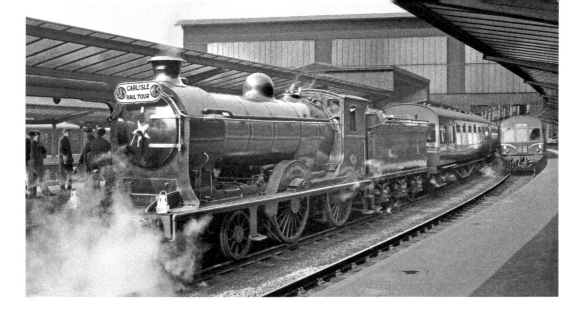

A Reid D34 Class 4-4-0, 256 *Glen Douglas* (formerly BR 62469), stands in Carlisle Citadel's bay Platform 5 on 6 April 1963, preparing to depart on the SLS/MLS 'Carlisle Railtour', the 1X85 13.05 departure around the various junctions and freight-avoiding lines in Carlisle, then on to Longtown, Canonbie and Langholm and back. Built in 1913 at Cowlair's works, there were a total of thirty-two North British 'Glens'. No. 62469 was repainted by BR into the NBR livery in 1959, one of four renowned locomotives of Scottish origin retained for special workings. (Author's Collection)

Seen through the 1958-installed steel and glass canopies, former East Coast Main Line stalwarts of different eras, A4 60009 *Union of South Africa* and HST power car 43013, are visitors to the West Coast Main Line (WCML) at Carlisle on Saturday 11 March 2013.

As an expectant traveller looks out for the arrival of his London-bound train, the crew of 45305 are equally anxious to get their rake of empty coaching stock out of the way of Platform 3 and shunted into the sidings at Carlisle station on 22 August 2012, prior to taking on water at Upperby.

Black 5 45305 stands at Platform 3 with the 1Z52 08.07 Lancaster–Carlisle 'Fellsman' charter on 22 August 2012, seen from the G&SWR and former Silloth and Waverley lines bay platforms 7 and 8 at Carlisle.

Driver Steve Chipperfield of the West Coast Railway Company has just checked the lamps on the front of Peppercorn K1 2-6-0, 62005, during a brief pause at Carlisle's Platform 4, the late-running 5Z66 12.17 Fort William–Carnforth empty stock move on 29 October 2012. It is 23.30 and the only other souls on the platform are members of the train crew, having alighted from the train. With a green aspect for the West Coast Main showing, it was just seconds before the K1 set off into the night on the final leg of its journey over Shap summit and down to the Lancashire coastal plain at Carnforth.

Black 5s 45407 and 44871 briefly pause at Carlisle on 30 October 2010, returning to Carnforth as the 5Z66 08.35 from Fort William, following completion of the 2010 season of the West Highland Extension 'Jacobite' services between Fort William and Mallaig.

No. 60163 *Tornado* is ready for the road, with safety valves lifting, heading the 1Z55 18.23 Glasgow Central–Crewe 'Caledonian Tornado' charter on 21 September 2011.

Peppercorn A1 Pacific 60163 *Tornado* caught at rest at Carlisle station on 21 September 2011, closely observed by an admiring bystander.

No. 60009 *Union of South Africa* gets underway from the Border City at 16.11, returning the 1Z56 'CME' to Worcester Foregate Street on 27 May 2013.

The last winter 'CME' of the season, with 60009 *Union of South Africa*, runs into Carlisle's Platform 3 on time; it is the 1Z86 07.06 from London Euston on Saturday 9 March 2013.

Driver Peter Kirk heads to the signal telephone to receive his instructions to undertake the stock shunting move from Platform 3 at Carlisle on 13 August 2012. Everyone on the station goes about their business, ignoring the presence of Black 5 45305 at the head of a twelve-coach train, giving a wonderful feeling of normality about this scene on Monday lunchtime.

It is 6.05 p.m. on 6 July 1964 and BR Standard Class 5 4-6-0, 73062, of Glasgow Polmadie, blasts away from Carlisle Citadel Platform 3 heading a Glasgow service. No. 73062 was withdrawn in June 1965 after a ridiculously short working life of around a decade, and was scrapped at Arnott Young, Old Kilpatrick, during October 1965. (Author's Collection)

A 1965 photograph of Britannia Pacific 70007 *Coeur-de-Lion* moving away from the coaching stock sidings at Carlisle Citadel. The first Britannia to be withdrawn from service, in June 1965, 70007 was quickly disposed of at Crewe works during the following month. (Author's Collection)

A delightful photographic view of the Citadel and Carlisle station was lost when the overhead line equipment was installed for the WCML electrification in the 1970s, but the clock is turned back by 46233 *Duchess of Sutherland* in the carriage sidings on 9 June 2012. The Citadel was originally built in 1543 to guard the approach to the city, but was rebuilt to the design of Robert Smirke in 1821 to house the assize courts.

Kingmoor's 9F 2-10-0, 92056, trundles through Shap station heading a loaded coal train for Carlisle yard in the summer of 1967. The 9F was withdrawn during November 1967 and scrapped at Motherwell Machinery & Scrap Co., Wishaw, in February 1968. The charming station at Shap, with its steeply gabled roof and glass veranda, did not survive for much longer, closing during the following year. (Author's Collection)

Nearing Shap summit, an unidentified Ivatt mixed-traffic Class 4MT 2-6-0 gives all she has to assist a morning freight service over the top of the summit in 1962. During this period there were three such locos allocated to Tebay shed: 43009, 43029 and 43035. (Author's Collection)

Britannia Pacific 70013 *Oliver Cromwell* is on the ascent towards Shap summit in charge of the twelve-coach 1Z86 07.09 London Euston–Carlisle 'CME', seen at Birkbeck Viaduct, near Low Scales, on 25 February 2012. Two crows playing on the wind accompanied the wonderful sight and sound of the Britannia climbing on the rising gradient of 1 in 75 past Greenholme, Scout Green and on to Shap summit.

In woebegone condition and with smokebox numberplate missing, Standard class 4MT 4-6-0 75030 moves off Tebay (12E) shed in the summer of 1967 to take up banking duties. A handful of Standard 4MTs were on Tebay's books during the last months of steam over Shap. No. 75030, transferred from Stoke shed in May 1967, was withdrawn when Tebay shed closed on 31 December 1967. After a short period of storage at Tebay and Kingmoor sheds it was dispatched for disposal at Arnott Young, Carmyle, and was dealt with during July 1968.

Britannia 70011 *Hotspur*, of Kingmoor depot, rounds the curve at Oxenholme in charge of an Up Class H freight at 10.30 a.m. on 7 March 1964. This former Norwich-allocated Pacific survived until the end of steam at Carlisle in December 1967 and was cut up at McWilliams, Shettleston, during April 1968. (Eric F. Bentley, Author's Collection)

A young sound recordist captures for posterity BR Crewe-built 5MT 44899, about to depart from Oxenholme, heading a Down evening passenger service on 7 June 1967. Oxenholme station, just south of Kendal, was opened in 1847 as Kendal Junction, and renamed Oxenholme in 1860. The Westmorland village grew around the station and is named after Oxenholme Farm. It does not have its own church so it is technically a hamlet. (Author's Collection)

BR Derby 1944-built Black 5 4-6-0, 44825, hurries south through the delightful Shap station during the summer of 1967. No. 44825 was withdrawn in the October, and the station closed shortly afterwards, on 1 July 1968, after giving over 120 years' service to the local community. (Author's Collection)

In the rain on Saturday 17 August 2013, 46233 *Duchess of Sutherland* heads the 1Z28 05.52 Crewe and Liverpool Lime Street–Carlisle 'CME' charter, coasting around the curve and crossing the River Petteril at Petteril Bridge Junction, the point that marks the northern extremity of the Settle & Carlisle route.

On 4 February 2012 and with a temperature of around 2°C in the pouring rain, 45305 gets into its stride heading the 1Z88 16.14 winter 'CME' departure to London Euston, and is seen here shortly after leaving the Border City, near Carleton, as darkness falls at around 4.30 p.m.

Britannia 70013 *Oliver Cromwell* storms confidently past Wharton Dykes in charge of the 1Z87 14.37 Carlisle–London Euston 'CME', climbing on towards Birkett Common, Mallerstang and Ais Gill summit on 25 February 2012.

No. 46233 *Duchess of Sutherland* in full cry on the 1 in 200 climb past the site of the former Crosby Garrett station, heading the 1Z87 14.28 Carlisle–London Euston 'CME' on 14 June 2012.

Curiously bearing an 'Irish Mail' headboard, 46115 *Scots Guardsman* rounds the curve through Kirkby Stephen station on a wet and blustery 23 June 2012, heading the seven-coach 1Z81 16.55 Carlisle–Scarborough private charter with ease. Before rebuilding, 'Royal Scots' rarely worked over the S&C, and when first allocated to Holbeck in 1943, crews previously familiar with 'Jubilees' were amazed at their potential over the 'Roof of England'.

BR Standard 8P Pacific 71000 *Duke of Gloucester* emerges from the rain into a brief spell of sunshine at Greengate, near Kirkby Stephen, heading the 1Z72 14.31 Carlisle–Gloucester return 'CME' on 21 May 2011. The legendary and solitary three-cylinder and Caprotti valve gear-fitted *Duke of Gloucester* was Britain's ultimate express passenger locomotive, intended to be the prototype of a standard design for fast and heavy expresses, but the modernisation plan saw an early end to its career after just eight years' BR service.

The ill-fated 'Cumbrian Ranger' charter, the 1Z46 14.40 Carlisle–Tyseley Warwick Road, with haulage by Gresley A4 4464 *Bittern* to Crewe, passes a milepost 298 feet above Armathwaite village on Saturday 15 March 2014. A temporary gauge restriction for A4 Class locomotives running through the platforms at Carlisle station forced the northbound leg between Carnforth and Carlisle to be diesel-hauled.

The 1Z81 16.50 Carlisle–York 'Waverley' charter on 22 August 2010 near Dry Beck, with BR Horwich-built Stanier 5MT 44932 in charge.

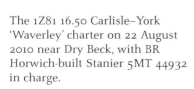

Viewed from Ais Gill signal box, with the distinctive shape of Wild Boar Fell looming in the background, 8F 48612 of Kingmoor shed breasts the summit (1,169 feet above sea level) heading Anhydrite empties from Widnes to Long Meg sidings in around 1963. The freight 2-8-0 lasted almost to the end of BR steam, being withdrawn from Newton Heath in June 1968, and cut up at Drapers, Hull, in November 1968. (Author's Collection)

No. 60009 *Union of South Africa* approaches the summit at Ais Gill, and the Cumbria/North Yorkshire county boundary, heading the southbound 1Z88 14.40 Carlisle–London Euston winter 'CME' on 16 February 2013.

Bathed in a brief flash of sunlight, 44932 approaches the Settle–Carlisle line summit at Ais Gill, heading the return 1Z73 15.45 Carlisle–York 'Waverley' charter on Sunday 2 September 2012.

No. 45699 *Galatea* slowly but positively climbs past Angrholm towards the summit of Ais Gill with the return southbound leg of the 1Z53 15.34 Carlisle–Lancaster 'Fellsman' on Wednesday 2 July 2014.

Hughes-Fowler Class 5 2-6-0, 42819, romps through Appleby West station in unforgiving rain on 23 May 1964, heading a northbound mixed freight for Carlisle yard. Not a great deal has changed in this scene today, except for the station renaming and new platform surfaces. At this time the Crab was Agecroft-allocated, but the class had been no stranger to Kingmoor, which had its own allocation up until the end of 1962. (Author's Collection)

Although few WCML depots had allocations of 'Crabs', Kingmoor had nineteen Hughes-Fowler examples on its books in the early 1960s up until the end of 1962, when all were either withdrawn or transferred away en masse. No. 42833 is seen at the north end of the shed yard in the ash disposal area in the summer of 1961. It was reallocated to Lancaster Green Ayre in June 1962 but withdrawn soon after, in the December. The Crab's inclined cylinders and curved running plates gave them a totally unique, almost period appearance. (Author's Collection)

Fowler Parallel 2-6-4 tank 42301 moves off Kingmoor shed in the mist and snow in December 1962. Introduced in 1927, the 4MT Class totalled 125, and they were the forerunners of the Stanier, Fairburn, and BR Standard 2-6-4 tanks. Hughes-Fowler Crab 2-6-0 42875 stands forlorn and withdrawn on an adjacent line and was eventually cut up during the following November. No. 42301 remained in service at Kingmoor until the following October and was cut up at Glasson Dock, Lancaster, in March 1964. (Author's Collection)

Doyen of the Ivatt Class 4 of mixed traffic 2-6-0s, 43000, rests near the entrance to the south end of Kingmoor shed on 30 August 1964. The BR 4MT 2-6-0 was based on the Ivatt 4MT design, the latter arguably having a more pleasing and functional appearance. (Author's Collection)

A close-up of the cab and token apparatus fitted to Ivatt 4MT 43139 at Kingmoor depot in 1967. A regular locomotive for the Langholm branch services, it was withdrawn in September 1967 and cut up at Motherwell Machinery & Scrap Co., Wishaw, during February 1968. (Author's Collection)

This is one of the author's earliest photographs, taken when fourteen with a Kodak Instamatic camera. Britannia 70042, formerly named *Lord Roberts*, is standing on the disposal road at Kingmoor on 1 April 1967. This once gleaming BR Standard Pacific was to see just a few more weeks in service, being withdrawn during the following month and subsequently despatched for scrap at McWilliam's scrapyard in Shettleston, near Glasgow.

The last members of the rebuilt Patriot and Royal Scot classes, 45530 *Sir Frank Ree*, divested of its chimney, and 46115 *Scots Guardsman* have been unceremoniously dumped at the south end of Kingmoor shed early in 1966, as if standing under the gallows awaiting their fate. Both were withdrawn from service at the end of 1965 – 45530 was cut up at Motherwell Machinery & Scrap Co., Wishaw, by July 1966, but 46115 happily survived the cutter's torch and is now a regular and reliable performer on the main line in the twenty-first century. The Patriot class was based on the chassis of the Royal Scot, combined with the boiler from large L&NWR Claughton Class 4-6-0, earning them the nickname 'Baby Scots'. There were fifty-two built between 1930 and 1934, and none survived, although the exciting project to build the new 45551 *The Unknown Warrior* will happily soon address this issue. (Author's Collection)

Despite its appearance, lurking inside the darkness of Kingmoor shed in May 1967, this Ivatt class 4MT 2-6-0, 43121, saw further service until final withdrawal came for it in the November. It was eventually disposed of by Motherwell Machinery & Scrap Co. at Wishaw in February 1968. (Author's Collection)

Although not lying in the county when this photograph was taken in 1963, Barrow-in-Furness is now very much part of Cumbria. The motive power depot also came under the control of the Carlisle District Locomotive Superintendent. Being slightly off the beaten track, photographs taken within the depot are rarely featured. Jinty 0-6-0 tank 47531 is keeping company with 4F 0-6-0, 44447, at the end of one of the ten shed roads. The 1874-built shed was located on the south side of the old Furness railway works, adjacent to Barrow docks. The Jinty was transferred to Warrington Dallam just before the shed closed to steam on 12 December 1966, but it later found its way back north to Kingmoor for a short period. (Author's Collection)

34

No. 47531 languishes at Kingmoor in the summer of 1966, by now withdrawn from service. It was despatched to T. W. Ward, Beighton, for scrap in March 1967. (Author's Collection)

Only ten 6P5F 'Clans' were built by BR between 1951 and 1952, the design being a lightweight version of the Britannia, with an identical chassis but smaller cylinders and lower boiler pressure, enabling a maximum axle load of 19 tons. The engines spent their lives operating daily over routes that encompassed Shap, Beattock, Ais Gill and the tortuous line to Stranraer. The five Polmadie-allocated class members were withdrawn en masse in December 1962; those at Kingmoor fared slightly better, being withdrawn between May 1965 and April 1966. No. 72007 *Clan Mackintosh* drifts down the grade from Moorcock Tunnel at Garsdale, heading the 16.25 Carlisle Citadel–Bradford service in 1965. (Author's Collection)

Britannia 70005, formerly named *John Milton*, adds to the smoky atmosphere at the south end of Kingmoor shed in July 1966. No. 70005 was originally one of the Great Eastern section Britannias, at home working the Liverpool Street–Norwich expresses, with speeds of up to 100 mph on some stretches of line. The Pacific lasted just one more year in service, being withdrawn during the following July and scrapped by Campbells of Airdrie in January 1968. (Charlie Cross)

A summer 1967 scene at the south end of Kingmoor shed, epitomising the filthy conditions of both locomotives and depot at this time. Visiting Black 5 44830 of Heaton Mersey awaits its next diagram, along with another unidentified Black 5, with double-chimney-fitted 9F 92223 simmering on the adjacent line, the 2-10-0 freight loco having barely seen nine years of service since construction. This 9F, the ultimate in BR steam locomotive design, was withdrawn from Carnforth in April 1968 and scrapped at Arnott Young, Dinsdale, during September 1968. At closure at the end of 1967, there were six 9F Class 2-10-0s allocated to Kingmoor. (Author's Collection)

No. 46115 *Scots Guardsman* was withdrawn in December 1965 and stored at Kingmoor until August 1966. The front end had received a fresh coat of paint prior to its move into preservation, initially to the Keighley & Worth Valley Railway and then on to the now closed Dinting Railway Centre. The 4-6-0 has a claim to fame in becoming the star locomotive in the famous 1936 GPO feature film *Night Mail* and was the last member of the class to be rebuilt by the LM&SR before nationalisation. It was the last of its class in BR service, withdrawn in December 1965. (Charlie Cross)

Britannia 70035, at one time proudly bearing the name *Rudyard Kipling* and a performer on Liverpool Street–Norwich crack expresses, at least demonstrates some dignity in this March 1967 photograph. An attempt to clean its unlined green livery adds a welcome splash of colour to the dismal surroundings of Kingmoor shed. Withdrawn with the mass cull of Britannia Pacifics upon the closure of Kingmoor in December 1967, it was scrapped at T. W. Ward, Inverkeithing, during April 1968. (Charlie Cross)

A nameless and unidentified Britannia Pacific faces 5MT 44915 of Lostock Hall depot, bearing chalked eyes on the smokebox door, at Kingmoor shed in July 1966. No. 44915 was withdrawn in December 1967. (Author's Collection)

Two Black 5s and three Britannias, including Nos 70023, formerly named *Venus*, and 70004, formerly named *William Shakespeare*, glint in the low evening sun at the south end of Kingmoor shed on 7 May 1966. (Mike Blenkinsop, Author's Collection)

A 1954 photograph of Polmadie-allocated rebuilt Royal Scot 46105 *Cameron Highlander* at Kingmoor. It is fully coaled and watered and, after turning upon moving off shed, will probably take a Glasgow passenger service forward from 'The Citadel'. (Author's Collection)

It is the end of the line for Royal Scot 46132 *The King's Regiment Liverpool*, still proudly bearing its nameplates and smokebox numberplate, awaiting its last journey to the breaker's yard. It is keeping company with Fowler 2-6-4 tank 42353, also withdrawn but still with a full bunker of coal, at the south end of the shed yard on a wet 28 July 1964. The 'Scot' was withdrawn during February 1964 and met its fate at West of Scotland Shipbreaking, Troon, during April 1965. (Author's Collection)

Kingmoor shed in the late 1960s with a line of stored locos, including 1947-built 5MT 44767 and two Britannias. The unique Black 5 out of a class of 842 examples, fitted with Stephenson link motion and Timken roller bearings, survived the cutter's torch, unlike the Britannias. (Author's Collection)

A May 1967 line up at the north end of Kingmoor shed, with BR Derby-built Black 5, 44816, an unidentified 9F 2-10-0 and another Black 5 all in steam, as well as Britannia 70018, formerly named *Flying Dutchman,* on the scrap line in the distance. No. 44816 lasted almost to the very end of BR steam, at Lostock Hall depot, and was finally cut up at Cohen's, Kettering, in early 1969. (Author's Collection)

BR Horwich-built Black 5 44672 stands in the evening sun at the south end of Kingmoor shed in 1967. Fitted with Skefco roller bearings on the driving coupled axle only, it was introduced to service in 1950. Transferred to Lostock Hall shed upon the closure of Kingmoor, she lasted until the March of 1968, and was cut up at Draper's of Hull during June 1968. (Author's Collection)

No. 70016 (formerly named *Ariel*) and Stockport's Black 5, 44940, rest between duties at the north end of Kingmoor shed on 16 July 1967. No. 44940 was one of the first batch of post-war Black 5s built at Horwich in 1945/46 and was withdrawn in March 1968. (Author's Collection)

The end of the line for so many locomotives that passed through Kingmoor depot, the scrap line alongside the coaling and ash disposal plants on the north-eastern perimeter, and conveniently the surreptitious access to the depot for many an enthusiast, the locos providing useful concealment from the shed offices! 8F 2-8-0 48287 and two BR Standard 9Fs, including 92018, devoid of its smoke deflectors, await their call to the breaker's yard in June 1967. The 8F was to remain on the scrap line until after the depot closed and was moved to Cashmore's at Great Bridge for scrap in April 1968. No. 92018, a sad remnant of BR's ultimate freight locomotive, was towed away to Motherwell Machinery & Scrap at Wishaw during July 1967. (Author's Collection)

Just shortly before its withdrawal from BR service in October 1967, Jubilee 45593 *Kolhapur* of Holbeck sits at the north end of Kingmoor shed in the company of Black 5, 45279, and Britannia 70011, formerly named *Hotspur,* on 20 August 1967. The Jubilee has clearly benefitted from some effective attention from members of the legendary Master Neverers Association (MNA) and it was one of the last two Jubilees to remain in service, the other being 45562 *Alberta* . (Author's Collection)

A fitter climbs aboard former Crosti-boilered Birkenhead 9F 92020, keeping company with an Ivatt 4MT 2-6-0 and a Britannia at Kingmoor shed in July 1966. The 9F remained in service until October 1967 and was disposed of at Buttigieg's scrapyard, Newport, during the summer of 1968. (Charlie Cross)

With just five months to go before its withdrawal from service, 47641, one of the 221 Fowler 3F 0-6-0 tanks, or 'Jinties' as they were affectionately known, brews up at Kingmoor shed in July 1966; a sombre scene which epitomises the condition that most locomotives were to be found in during this period. (Charlie Cross)

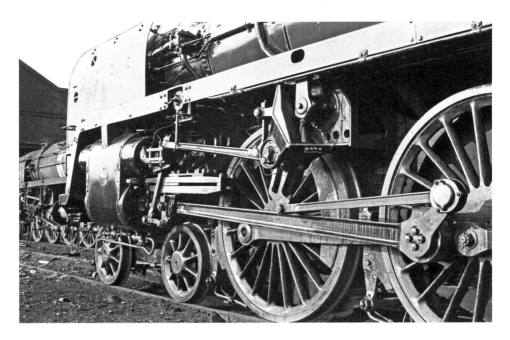

Britannia 70013 *Oliver Cromwell* basks in the evening sun at Kingmoor depot in June 1967, three months after its full overhaul and repainting at Crewe Works. (Author's Collection)

Immaculate 70013 *Oliver Cromwell*, its nameplates removed for safe keeping, stands at the south end of Kingmoor shed in the company of a workaday 9F in June 1967. In addition to the well-known 1T57 'Fifteen Guinea Special' duty of 11 August 1968, working into Carlisle from Manchester Victoria, 70013 had in fact worked the last BR steam-hauled passenger working out of Carlisle before the end of regular steam duties there, a 451-ton football supporters' special to Blackpool. Leaving Carlisle at 09.45 on Boxing Day 1967, it proved to be the last BR steam-hauled passenger service over Shap summit. (Author's Collection)

An evening visit by a rail fan club to Kingmoor depot in the summer of 1966 produced visiting BR Standard class 5 4-6-0 73100 of Corkerhill depot, and local residents, Black 5, 44900, and Britannia 70034, formerly named *Thomas Hardy*. (Author's Collection)

A Speke Junction-allocated 9F and a line of Black 5s, with Armstrong Whitworth-built 45388 at the head, stand at the north end of Carlisle Kingmoor depot alongside the ash disposal point in 1966. (Mike Blenkinsop, Author's Collection)

Britannia 70028 (formerly named *Royal Star*) inside Kingmoor running shed in 1966, with two Stanier 5MTs outside, the nearest being 45481. Originally one of Cardiff Canton's Britannias, she once worked expresses such as 'The Red Dragon' and 'The Capitals United'. (Author's Collection)

BR Standard Class 5, 73107, tows V2 class 2-6-2 60816 and Ivatt 4MT 43049 out of the shed at Kingmoor on 19 March 1966. With the closure of Carlisle Canal shed in June 1963, V2s became regular visitors to Kingmoor. (Mike Blenkinsop, Author's Collection)

No. 70040 *Clive of India* has reached the end of the road; this is a summer 1967 photograph of it outside the workshops at Kingmoor, withdrawn from service and with its pony truck removed. It was to remain there until the November, then despatched to McWilliams' scrapyard, Shettleston, and scrapped during December 1967. (Author's Collection)

Seen on a sunny evening in June 1967 and before painted names were applied to its smoke deflectors, 70013 *Oliver Cromwell* is in steam at Kingmoor depot, the pride of Carlisle of the period. The Britannia was the last steam locomotive to receive a major overhaul at Crewe Works and emerged in February 1967. (Author's Collection)

A location much favoured by the author for observing the action in the 1960s was the Waverley overbridge at the south end of Carlisle New Yard, with its bicycle shed opposite the Power Signal Box providing shelter from the rain when required. Britannia 70040 *Clive of India* is well into her stride heading a Perth train in 1966. The driver acknowledges the photographer and an enthusiast is enjoying the thrash in the vestibule of the leading coach. The former 'front line' Norwich-allocated Pacific was withdrawn in April 1967 and cut up at McWilliams' scrapyard, Shettleston, in December 1967. (Author's Collection)

A mid-1960s photograph of Stourton's 8F 48084 heading a well-laden coal train around the short-lived spur at Kingmoor Junction, which provided a freight link from Stainton Junction to Carlisle New Yard. A smoky Kingmoor shed presents a busy scene behind. The 8F was withdrawn from service at Royston shed in November 1967. (Author's Collection)

On Saturday 21 April 2012, 46115 *Scots Guardsman* crosses the Mossband viaduct near Gretna, heading the 5Z26 Carnforth to Edinburgh Joppa positioning move for the 'Great Britain V' railtour. Recent flooding of the River Esk had left debris at the highest tide line.

No. 46115 *Scots Guardsman* crosses the splendid sandstone two-arch bridge over the River Sark at Gretna, which marks the border between England and Scotland, working a positioning diagram from Carnforth to Thornton Junction on 29 June 2012. The seventy-one class members, all named after Regiments, did sterling service on the West Coast Main Line. No. 46115, as the last survivor, ended her BR career on local freight trips, a regular turn being between Carlisle and Mossend.

Armstrong Whitworth-built Black 5, 45312, coasts by Etterby Junction with a freight for the New Yard on 13 June 1967. A Warrington locomotive, it was withdrawn from service at Bolton in May 1968 and despatched for cutting up at Cohen's, Kettering, in early 1969. The loco's top lamp bracket has been moved down to the left-hand side of the smokebox to avoid danger from the 25kV electric catenary. (Author's Collection)

Another wet and grey day at Carlisle sets the scene viewed from Etterby Bridge in July 1966. An unidentified 9F and a Britannia stand on the Eden river bridge. Both would appear to have full tenders, so are presumably moving off shed to take up diagrams. The sidings shunted by William Bagnall fireless locos, with a coal conveyor feeding Willowholme power station, can be seen on the right and the much-favoured official railway staff walking route to town is also visible. The closed footbridge survived until 2011, despite the path having been removed many decades previously. (Mike Blenkinsop, Author's Collection)

A distinguished survivor of the time, 46115 *Scots Guardsman* crosses the Eden river bridge at Etterby Junction on the Down goods line, heading a 3X35 empty coaching stock move north from Carlisle. The presence of the yellow cabside stripe on the loco would suggest a post-August 1964 photograph. The goods lines on the right from Port Carlisle Junction to Etterby were installed by the LM&SR in 1942 to cope with the additional traffic generated by the Second World War. (Author's Collection)

With a cold north-easterly wind blowing and at 10 degrees Celsius, former Upperby shed resident Crewe 1944-built Black 5, 44871, in what can best described as workaday condition, climbs away from the Border City, crosses the Eden river and forges north heading the 5Z49 Carnforth–Fort William empty stock move at 7.25 a.m. on 2 June 2012, the locomotive bearing 'The Mancunian' headboard.

The bleak countryside and steep gradients attracted the most eminent photographers of the day to the area. On what possibly proved to be one of its last diagrams before withdrawal, Jubilee 45584 *North West Frontier* is in charge of a 1S54 Down passenger and is struggling to gain momentum on the wet rails on the climb to the summit, passing Shap Wells in the summer of 1964. Jubilees handled most of the passenger traffic between Lancashire and Scotland in the 1950s and early 1960s. Withdrawn from Kingmoor depot in September 1964, shortly after transfer from Blackpool, the Jubilee was cut up at Hughes Bolkows Ltd, North Blyth, in January 1965 after a period of three months' storage at Kingmoor. (A. E. Durrant, Author's Collection)

The spectacle of a 1933-built Stanier LMS Princess Royal class Pacific in full cry on the climb past Shap Wells, the very duty it was designed to undertake, is an amazing experience in the twenty-first century. No. 6201 *Princess Elizabeth* is seen on the 1 in 75 climb from Scout Green to Shap summit in charge of the 1Z90 06.30 Tyseley Warwick Road–Carlisle 'Cumbrian Mountaineer' on Saturday 17 November 2012. The class of thirteen locomotives was built from 1933 to handle the heaviest Anglo-Scottish expresses and 6201 is the record holder for the fastest runs by steam between London and Glasgow during the tests of 1936.

The 1Z86 06.26 London Euston–Carlisle 'CME' between Greenholme and Scout Green on 14 June 2012, with 46233 *Duchess of Sutherland* climbing the four miles of 1 in 75 to Shap summit in fine fettle and doing exactly what it was originally designed for with confidence.

This 1L00 Warrington–Carlisle morning stopping train didn't present any problems for Vulcan Foundry-built Black 5, 45039 of Edge Hill shed, as it storms the grade to the summit, passing Shap Wells in March 1963. The Shap Wells Hotel nearby was built in 1850 after the discovery that the local water had beneficial properties and the large number of red squirrels to be found living in the adjoining woodland appear to thrive very well on it. (Author's Collection)

No. 46115 *Scots Guardsman* performing impeccably on the climb to Shap summit at Low Scales, heading the 1Z86 07.08 London Euston–Carlisle winter 'CME' on 15 February 2014. No. 46115 was rebuilt in 1947 with a type 2A taper boiler, new cylinders and double chimney. It was also the only Royal Scot to be fitted with smoke deflectors while still in LMS livery.

No. 45388 of Wigan Springs Branch rounds the curve at Greenholme on the ascent to Scout Green and Shap summit, heading a train of cattle wagons for Carlisle in 1962. The Black 5, built between 1934 and 1950, was probably the best all-round steam design built in the UK and they could be entrusted to any task put before them. No. 45388 lasted right up until the end of BR steam, latterly at Lostock Hall, and was one of the last three non-preserved BR steam locomotives to leave a BR shed, consigned to scrap at Draper's of Hull in April 1969. (Author's Collection)

After an overnight ground frost and with a cold north-easterly wind blowing on 22 March 1963, Crewe-built Black 5,44883, with assistance from the rear rendered by an unidentified Fairburn 2-6-4 tank doing its fair share of work, slogs up the grade past Scout Green with a partly fitted Class E express freight for Carlisle, which included insulated meat containers. The assistance of one of Tebay's banking locomotives would often be required for the 1 in 75 climb to Shap summit. (Author's Collection)

In blizzard conditions, Gresley A4 Pacific 60009 *Union of South Africa* storms the grade towards Shap summit at Scout Green, adjacent to where the signal box once stood. A truly impressive performance by both man and machine, pitted against the terrain and the elements on the 1Z86 07.06 London Euston–Carlisle winter 'CME' of 23 February 2013.

No. 46115 *Scots Guardsman* slowly but positively tackles the grade to Shap Summit past Shap Wells, having just been looped at Tebay due to an issue with a preceding Virgin 'Pendolino'. This was the 1Z86 07.08 London Euston–Carlisle winter 'CME' on 1 March 2014.

During a bright spot on a typically wet day in the Lune Gorge at Dillicar, Princess Coronation Pacific 46233 *Duchess of Sutherland* heads the 1Z75 05.16 Sheffield–Perth '75th Anniversary Special' charter on Friday 6 September 2013.

No. 60009 *Union of South Africa* hurries across Birkbeck viaduct over the infant River Lune at Low Scales Farm, heading the 1Z86 07.06 London Euston–Carlisle winter 'CME' on 16 February 2013.

No. 60009 *Union of South Africa* puts in another outstanding performance on the climb to Shap summit, heading the 1Z63 09.47 Bangor–Hexham 'Cathedrals Explorer' railtour on 13 May 2013. A combination of an unusually late path for steam over Shap, a booked time at the summit of 15.56 and the strong south-westerly wind made Salterwath near Scout Green an ideal choice of location for this train. The 5 degrees Celsius temperature also ensured a satisfactory amount of exhaust, although the strong westerly wind quickly tugged it away across the fells.

Stanier Princess Royal class 4-6-2 6201 *Princess Elizabeth* was an impressive sight as it made the steady climb to Shap summit at Stonygill with the 1Z90 06.30 Tyseley Warwick Road–Carlisle 'Cumbrian Mountaineer' charter on 17 November 2012.

The beach bungalows between Nethertown and Braystones were originally sited along the coastline here by railway navvies working on the building of the Furness Railway in the late nineteenth century. The railway was constructed right through the massive sand dunes along this exposed stretch of the Cumbrian coast. With a sea fret adding to the atmosphere of this unequalled location in England, 46115 *Scots Guardsman* climbs through Braystones station heading the 1Z60 Grange-over-Sands–Edinburgh 'Great Britain VII' charter on 29 April 2014, evoking memories of some London Euston–Workington Main services which were entrusted to Royal Scot class 4-6-0s in the 1950s.

On an appropriately murky morning, and running early at around 8.30 a.m. on 21 April 2012, Carnforth's Black 5, 45305, scurries across the Eden River Bridge at Kingmoor, working the 5Z33 08.20 Carnforth–Edinburgh Joppa positioning move. The Black 5 was allocated to Workington shed during the early part of the 1960s and was undoubtedly a regular visitor to Carlisle and Kingmoor shed during its career, although it finished its last days in BR service working from Lostock Hall.

Gresley K4, 61994 *The Great Marquess*, drifts past the former Kingmoor steam MPD site and passes beneath the Etterby Bridge, heading the 1Z94 08.45 Barnhill–Preston 'West Highlander' on a drizzly Tuesday 24 September 2013. The name Kingmoor is derived from the king's former hunting moor north of Carlisle. In 1332–33 Edward III granted to the citizens of Carlisle 'liberties and customs – the right to common pasture for all kinds of animals at all times of the year upon the King's moor, with right to dig turfs there and carry them away at will'. By tradition, Ascension Day races were held at Kingmoor, the earliest reference to Kingmoor Races being in 1619.

Class 7P 4-6-0, 46115 *Scots Guardsman*, climbing at a positive pace at Lambrigg, heading the 1Z86 07.08 London Euston–Carlisle winter 'CME' on 1 March 2014.

No. 46115 *Scots Guardsman*, heading the 1Z60 Grange-over-Sands–Edinburgh 'Great Britain VII' charter on 29 April 2014, passes the former Maryport & Carlisle station of Curthwaite. The listed 1843-built M&C water tower and overbridge are of particular note at this former station, which was closed to regular passenger services in 1950.

Carnforth-based Black 5, 44932, has just arrived at Carlisle and the loco is being prepared to take forward 'The Fylde Coast Express' from the Brush Type 4 bringing the train from Kilmarnock. The Victorian cast-iron structure of the Rome Street Gasworks provides a suitable period backdrop with the clouds in the western sky being illuminated by the rising sun. The sky shimmers against the heat from the smokebox of the Black 5 as its fire is built up and the tender is topped up with water for the journey ahead over the 'Midland Route' and on to Preston and Blackpool. The resident urban seagulls complete the dawn scene at 5.45 a.m. on 22 May 2013.

Around fifty years ago Thompson B1 Class 4-6-0s were routinely seen around the Border City and even worked Tyne Valley service trains for a period of time, therefore what a great moment it was to see this dramatic departure from Carlisle with 61264 piloting Black 5, 45407, seen approaching Petteril Bridge Junction heading the return 1Z49 14.44 Carlisle–Manchester Victoria winter 'CME' on 25 January 2014. In March 1959 Carlisle Canal shed had six B1s at its disposal for mixed traffic duties.

Black 5 No. 44932 shuffles away to Upperby sidings for servicing and turning while the tables are being set in the dining cars of the 'CME' of 12 August 2012, in preparation for its return south later in the afternoon. A Virgin Pendolino stands in the adjacent platform.

Pure opulence of a bygone era at Carlisle station on 29 July 2012, personified by this fine BR Mk 1 Pullman first-class parlour car, *Emerald*, and the 'steam age' wall murals glimpsed through the carriage windows.

Returning from its historic visit to Edinburgh, Collett 1936-built Castle Class 4-6-0 5043 *Earl of Mount Edgecumbe* pauses at Carlisle, returning as the 1Z24 16.11 departure to Tyseley Warwick Road on 28 May 2012.

A steam charter train arrival under the overall roof at Carlisle is perfectly illuminated on a sunny midsummer's day. Such a spectacle adds further splendour to an already delightfully period survivor on our national network, despite considerations for modern-day passenger safety and information. Passengers waiting for a service train are clearly amazed at the arrival of 61994 *The Great Marquess* as it wheezes into platform 3 after what was understood to be a troublesome run for the footplate crew of the 1Z52 07.08 Lancaster–Carlisle 'Fellsman' charter via the S&C on 19 June 2013.

Drivers Kevin Treeby and Mick Rawlings on the footplate of Bulleid un-rebuilt Battle of Britain light Pacific 34067 *Tangmere* take a well-earned breather after arriving at Carlisle at the head of the 1Z86 06.26 London Euston–Carlisle 'CME' on 12 April 2012.

To the delight of three passengers wielding camera phones and compact cameras, 8F 48151, bearing a most pleasant patina, clanks into Carlisle on 25 August 2013, in a manner so synonymous with these mundane but reliable freight engines. This locomotive has certainly proved its worth in recent years, rescued as a hulk from Woodham's Barry scrapyard in November 1975. Its last two years of BR service were seen out from Northwich shed, being withdrawn from service in the March of 1968.

Almost four years since Carlisle station was last witness to the arrival of an un-rebuilt Bulleid Battle of Britain Pacific, 34067 *Tangmere* looked stunning rolling into Platform 3 heading the 1Z86 'CME' from London Euston on 12 April 2012.

Stanier's 'workhorse' 8F class 2-8-0 No. 48151 slogs up the final few yards to the summit of Ais Gill, heading the return 'Cathedrals Express' charter from Carlisle to Radlett on Thursday 5 June 2014.

The fireman of Bulleid Battle of Britain Pacific 34067 *Tangmere* prepares to put another round in the firebox shortly after the 1Z86 'CME' arrival at Carlisle from London Euston on 12 April 2012.

Class 8F No. 48151, with Driver Gordon Hodgson in charge, has just arrived at Carlisle heading the 1Z70 09.12 York–Carlisle 'Waverley' charter on 25 August 2013, and an amusing incident has clearly attracted the footplate crew and a bystander alike.

As held by Greek mythology, 'it is not carved of ivory, or milk-white', but happily the locomotive did come to life again. One time Barry scrapyard wreck 45699 *Galatea* runs through Platform 3 at Carlisle, heading the 1Z45 07.17 Tyseley Warwick Road–Carlisle 'Cumbrian Jubilee' charter on 9 November 2013.

The fireman of Peppercorn A1 Pacific 60163 *Tornado* adjusts his cap after stepping down from the footplate and there's no doubting the hard work that this member of the footplate crew has put in during his 140-mile roster from Crewe, the 1Z60 06.58 London Euston–Carlisle 'Cathedrals Express' on 31 May 2012.

Driver Steve Chipperfield of 5MT 45305 enjoys a chat and a well-earned 'cuppa' shortly after arrival at Carlisle on 13 August 2012, with the 1Z80 'Mersey Moorlander' charter, originating from Crewe at 06.06 and Liverpool Lime Street at 07.19.

Driver Brian Grierson of the return 1Z62 15.45 Carlisle–London Euston 'Cathedrals Express' via the Settle–Carlisle line on 31 May 2012 waits for the right away in charge of A1 Pacific 60163 *Tornado*. Brian was the last steam driver at DB Schenker's Carlisle Kingmoor depot, retiring on 28 March 2014 with fifty years' service to his name. Known simply as No. 1, he was the last of his kind, having seen in the end of mainline steam on BR just as his career was starting as a fireman in Carlisle.

Driver Peter Kirk and Neil 'Bubbles' Henderson, who was the fireman of the northbound 1Z80 'Mersey Moorlander' charter from Crewe at 06.06 and Liverpool Lime Street at 07.19 with Black 5 No. 45305 from Preston, take a well-earned breather at Carlisle on 13 August 2012.

A loco footplate crew member from Black 5 No. 45305, at the head of the Liverpool–Carlisle 'Mersey Moorlander' summer holiday special, goes about some important business with his 'tea mashing can' shortly after arrival at Carlisle station in British summertime weather on 6 August 2012.

Driver Cubits of 'The Hadrian' charter on 2 June 2012 exchanges banter on the footplate of 46233 *Duchess of Sutherland* before the 16.01 booked 1Z79 departure from Carlisle via the Tyne Valley line to York, where diesel traction was to take over to Leicester. Appropriately for the route and former BR region, Driver Cubits has a tangerine totem enamel badge in his grease top cap.

Stanier Jubilee 45699 *Galatea* approaches the summit of Ais Gill in typically overcast and blustery conditions on Thursday 3 July 2014, heading the return 'Cathedrals Express' charter from Carlisle to Oxford.

Stanier Jubilee 45675 *Hardy* heads the 1M32 Glasgow–Leeds and St Pancras away from Carlisle Citadel on 20 August 1966, receiving much attention from photographers on Platform 3. Black 5 No. 45363 stands in one of the centre roads, waiting to take over a following service. The Jubilee was withdrawn in July 1967, then moved to Wakefield depot for storage before being despatched for cutting up at Cashmore's, Great Bridge, in November 1967. (Author's Collection)

Bystanders are riveted to the platform at the splendour of Gresley A4 Pacific 60009 *Union of South Africa* beneath the overall roof of Carlisle with the 1Z60 'The Hadrian' charter from Skipton and York on Saturday 27 October 2012.

Fortunately the photographer was more focused on capturing the scale of the dilapidated overall roof and southern end screen at Carlisle Citadel before its disappearance rather than including the whole of Perth's BR Horwich-built Black 5 No. 44960 in the frame. A mid-1950s photograph, this was just before 1956–57 when BR removed the badly rotted parts of the roof, including the end screens, cutting the roof back and installing new end screens as we see today. Citadel station's iron and glazed roof is supported on a series of hooped trusses, only some of which remain today, for they used to cover the lines west beyond platform 1. The roof dates from 1873–76, when the complex was enlarged to cope with traffic brought by the opening of the S&C. Built to serve the Lancaster & Carlisle line and the Caledonian Railway, the station was eventually to serve seven different railway companies. Despite its poor condition here, it still retained some of its original grandeur. (Author's Collection)

Seen through the footbridge latticework, Jubilee 45699 *Galatea* rolls into Platform 3 at
Carlisle on 23 October 2013, heading the 1Z86 06.26 London Euston–Carlisle 'CME'.

Ivatt 2MT 2-6-2 tank 41264 stands with vans in platform 1 while acting as station pilot
at the south end of Carlisle Citadel station on 25 June 1966. (Mike Blenkinsop, Author's
Collection)

A mid-1950s scene taken from the arched footbridge in the centre of Carlisle Citadel station, with Crewe North-allocated Princess Coronation Class Pacific 46233 *Duchess of Sutherland* coasting into Platform 3, heading an afternoon Glasgow Central–London Euston express. The intensive station activity going on around is fascinating to study. (Author's Collection)

Patriot 4-6-0, 45513, one of a handful of unnamed members of the class, impatiently waiting for its next duty south, stands beneath the fine overall roof at 'The Citadel' in the early 1950s. Despite the Midland influence in their design, the Patriots were almost exclusively employed on the West Coast routes, about half the class being used on the Lancaster & Carlisle section on all types of train. No. 45513 had been rebuilt in 1932 from L&NWR Claughton No. 2416, which curiously had also remained nameless throughout its life. (Author's Collection)

Unnamed and un-rebuilt Patriot 45513 with its main driving rods off inside Carlisle Upperby roundhouse on 7 May 1960. Following transfer to Liverpool Edge Hill, the 4-6-0 was withdrawn in late 1961 and cut up at Crewe Works in October 1962. Upperby roundhouse was constructed shortly after nationalisation and was closed to steam on 31 December 1966. Diesel locos were serviced for a brief period before the roundhouse was given over to the Civil Engineering Dept., until final demolition came in 1978. The carriage shed, of L&NWR/LM&SR origin, ironically still remains to this day, albeit in a derelict condition. (Author's Collection)

Push-pull fitted Ivatt 2MT 2-6-2 tank 41217 inside the roundhouse at Upperby in May 1966. The class was introduced between 1946 and 1952 and were based on the LMS Stanier 2-6-2T which was, in turn, based on the LMS Fowler 2-6-2T. No. 41217, built at Crewe in 1948, had spent time at Widnes, Warrington and Southport before being transferred to Upperby in February 1965 and remained in service at Carlisle until December 1966, being scrapped at McWilliams, Shettleston, during September 1967. (Author's Collection)

Stanier Royal Scot 46134 *The Cheshire Regiment* inside the roundhouse at Carlisle Upperby on 6 October 1962. Withdrawn just two months later, it was cut up at Crewe Works in April 1963. Upperby was one of the principal sheds on the WCML and at one time had representatives from all the largest classes of West Coast express locomotives on its books. (Author's Collection)

Armstrong Whitworth 1935-built Black 5 No. 45126 of Kingmoor and Royal Scot 46123 *Royal Irish Fusilier* at Carlisle Upperby, probably early in 1963. No. 46123 was stored at Upperby from October 1962 until March 1963, and then towed to Crewe Works for cutting up, which took place during the April. The 'dummy' overhead electrification equipment used for gauging and training purposes is noteworthy.

A late 1962 view of Upperby depot yard, seen from the footbridge providing access from Hasell Street, located on the western side of the WCML. Maroon-liveried Princess Class 8P Pacific 46200 *The Princess Royal* is in storage and it would eventually be scrapped at Connell's, Coatbridge, in September 1964 after a failed preservation bid. Also to be seen in the yard are Fowler 4F 0-6-0 44009, Jubilee 45703 *Thunderer* and a Royal Scot Class 4-6-0. English Electric Type 4 diesels are also stabled in the roads immediately adjacent the carriage shed. (Author's Collection)

Observed by a rail fan on an official shed visit to Upperby, probably in the late 1950s, un-rebuilt Royal Scot 46163 *Civil Service Rifleman* of Crewe North shunts a Black 5. The ubiquitous Army surplus store respirator bag was a 'must have' accessory of the time for carrying Ian Allan ABCs, shed directory, notebook and sandwiches. (Author's Collection)

Carlisle Upperby Motive Power Depot, probably in 1965, with Stanier Black 5 No. 45052 present along with other members of the class, a Royal Scot Class, 'Jinty' and Ivatt 2MT, as well as the infiltrating English Electric Type 4 diesels. The LMS-built enginemen's lodgings, seen standing prominent on Gallows Hill beyond the Black 5 and diesels, still survives and has been converted into the Swallow Hilltop Hotel. The roundhouse, built in 1948, had thirty-two roads centred on a 70-foot turntable and was eventually demolished in 1978/79. (Author's Collection)

Un-rebuilt LMS Royal Scot 6162 *Queen's Westminster Rifleman* passes the Hasell Street footbridge and Upperby depot heading a Down express in the late 1940s. (Author's Collection)

Britannia 70031, formerly named *Byron*, in deplorable external condition outside the roundhouse at Upperby in July 1966. The Pacific was to remain in service, latterly at Kingmoor, until November 1967. (Charlie Cross)

It is hard to believe when you look at the photographs of 70013 *Oliver Cromwell* elsewhere in this book, taken less than one year later, that this is the same locomotive. What was to be the last BR express passenger locomotive, and shortly before its full overhaul at Crewe Works, stands outside the roundhouse at Upperby in July 1966, in the company of sister loco 70031. (Charlie Cross)

Above: Britannia 70013 *Oliver Cromwell,* in woebegone condition, moves off Upperby shed in July 1966. (Charlie Cross)

Right: One of the official walking routes between the former roundhouse/shed yard and the carriage/diesel depot at Upperby, through the ornate doorway, is still showing some splendour despite its derelict condition when photographed in 2013.

In the closing days of BR steam action on the WCML, and just days from its withdrawal from service, Britannia Pacific 70023 (formerly named *Venus*) slams through the rock cutting near Hay Fell on Grayrigg bank near Lambrigg, heading a northbound afternoon freight on Saturday 9 December 1967. (Eric F. Bentley, Author's Collection)

A juxtaposition of old and new in a number of respects: K1 Class 2-6-0, 61994 *The Great Marquess*, crosses Crown Street, Carlisle, with the empty stock for the return 1Z53 15.34 Carlisle–Lancaster 'Fellsman' charter on Wednesday 12 June 2013, catching a mother and child totally by surprise.

On Monday 20 August 2012 and seen from Hasell Street, Black 5 No. 45305 passes the site of the former Carlisle Upperby shed, overshadowed by the former Engineman's Lodgings standing on what used to be called Gallows Hill. The lodgings were subsequently transformed into a city hotel, now called the Swallow Hilltop. The footbridge leading to the locomotive shed used to be sited at the end of this road. Two residents view the passing of the southbound steam-hauled 'Mersey Moorlander' charter, hauled by a loco that no doubt visited Upperby shed many times during its BR service.

With the Crown Street under-bridge lights still on, it is just after 5.30 a.m. on Wednesday 22 May 2013 as slightly grubby Black 5 No. 44932 arrives at Carlisle from Carnforth to take over the 'Fylde Coast Express' two hours later. With cylinder drain cocks open and the exhaust filling Crown Street and station with smoke, Driver Gordon Hodgson opens the regulator and eases forward to Wapping Sidings in order to take on water and build up the fire in preparation for the journey ahead over the 'Midland Route' to Hellifield and onward to Preston and Blackpool. It was hard to believe that this scene was taking place in the twenty-first century!

A seventy-five-year-old Black 5 No. 45305 sets back into Carlisle station, virtually un-noticed apart from by one small boy, as busy city life goes on around Crown Street on Wednesday 22 August 2012.

On 25 June 1966 Wigan Spring's branch's Stanier Black 5 4-6-0, 44873, blows off impatiently in the bay platform 5, with an English Electric Type 3 and a 'Derby Lightweight' diesel multiple unit in the two centre roads. (Mike Blenkinsop, Author's Collection)

Ivatt 2MT 2-6-2 tank 41222, one of Citadel's regular station pilots during 1966, crosses Crown Street while performing pilot duties around the southern end of the station. The Ivatt 2MT tanks displaced the 'Jinty' 0-6-0 tanks from these duties during 1966. (Author's Collection)

With Crown Street Goods Depot dominating the background scene, 'Jinty' 3F 0-6-0 tank 47408 of Upperby depot is caught during a quiet spell between its station pilot duties at the south end of Carlisle Citadel station on 9 August 1961. Station pilots exhibited one red tail lamp and one red head lamp when operating within the limits of the station and yards. (Author's Collection)

'Jinty' 47515 of Kingmoor depot, on the Citadel north end station pilot duty, stands alongside the station's unmistakable west retaining screen wall, possibly in early 1964. Withdrawn from Kingmoor in July 1964, it was cut up at Campbell's, Airdrie, in January 1965. (Author's Collection)

'Jinty' 47326 attached to a rake of ex-LMS parcels vans at the north end of Platform 1 in the rain on 25 May 1965. The lack of shedcode plate would suggest its recent transfer from Upperby to Kingmoor depot in Carlisle. (Author's Collection)

A 1966 photograph of 'Jinty' 47415 leaving platform 2, the Maryport & Carlisle bay, while employed on station pilot duties at Carlisle Citadel in 1966. Britannia 70052 *Firth of Tay* can just be seen waiting to depart from the bay platform serving the Newcastle/S&C lines and an Ivatt 2-6-0 stands with vans in platform 1. (Alan Orchard, Author's Collection)

No. 46233 *Duchess of Sutherland* eases out of Carlisle on the approach to London Road Junction, heading the 1Z79 16.01 Carlisle–Leicester 'The Hadrian' charter via the Tyne Valley line on 2 June 2012.

In the harsh backlit midday sun, 46115 *Scots Guardsman* makes the sharp 'switchback' climb through the sand dunes up to Nethertown, heading the 1Z60 Grange-over-Sands–Edinburgh leg of the 'Great Britain VII' charter on 29 April 2014.

Hot on the heals of another steam-hauled departure that passed over the bridge just five minutes before, hauled by *The Great Marquess*, Royal Scot 46115 *Scots Guardsman* heads the 5Z48 09.37 Thornton Yard–Carnforth Depot empty stock across the 'freight avoiding lines' at Carlisle Bog Junction on Tuesday 24 September 2013. Bog Junction originally provided connections for exchange of traffic between the NER and LNWR goods lines and the Maryport & Carlisle/Cumbrian Coast.

For a few miles along the WCML between Gretna Junction and Carlisle station, Stanier Pacific 46233 *Duchess of Sutherland* returned to its 1950s/60s familiar ground when allocated to Crewe North shed. The 8P Pacific passes the site of the erstwhile Carlisle Kingmoor motive power depot and under Etterby Bridge, on the northern outskirts of Carlisle. This was the 1Z53 09.37 Glasgow Barnhill–Carlisle and Preston 'Great Britain V' railtour on 26 April 2012.

The shadows are dancing across the landscape, as if to intentionally tease the photographer, as 46233 *Duchess of Sutherland* is hard into the 1 in 75 climb to Shap summit, passing Shap Wells with the 1Z86 06.26 London Euston–Carlisle 'CME' on 22 May 2013.

Despite not being blessed with the sun for the departing 1Z23 16.14 Carlisle–Manchester Victoria and Crewe 'CME' on 7 September 2013, 60009 *Union of South Africa* still picks up some of the light from the sky as it storms away from the Border City at St Ninians.

Evoking memories of Leeds Holbeck-based former LNER Pacifics working on the 'Midland Route' in the early 1960s, Peppercorn A1 60163 *Tornado* rounds the curve past Howe & Co. Siding and tackles the 1 in 132 incline with ease towards Low Cotehill, early into its return journey to Preston, with the 1Z62 15.45 Carlisle–London Euston 'Cathedrals Express' on 31 May 2012.

A timeless scene as the fireman takes a rest from the toil of shovelling coal and watches the road ahead as 8F 48151 plods up the 1 in 166 grade towards Baron Wood, high above Kirkoswald and the Eden Valley, heading the 1Z52 08.07 Lancaster–Carlisle 'Fellsman' charter on 25 July 2012. There were 852 8Fs, built by the 'Big Four' during the Second World War, but also constructed by Beyer Peacock, North British and the Vulcan Foundry, a number ending their days in Egypt, Iraq, Italy, Israel, Persia and Turkey.

BR Standard 8P Pacific 71000 *Duke of Gloucester* rounds the curve into Wigton heading the 1Z26 07.08 Crewe–Carlisle 'Cumbrian Coast Explorer' along the Maryport & Carlisle section of the Cumbrian Coast line on 3 July 2010. No. 71000 was the last BR express passenger locomotive to be built in 1954, but was withdrawn from BR service as early as November 1962.

It is a perfect late spring morning alongside Morecambe Bay as Driver Gordon Hodgson faultlessly gets 46115 *Scots Guardsman* into her stride upon departure from Grange-over-Sands, at Ca Lane crossing, heading the 1Z60 'Great Britain VII' charter to Edinburgh on 29 April 2014.

The early 1970s witnessed a number of Carnforth–Sellafield steam-hauled excursions with locomotives based at the Steamtown museum, the former 10A motive power depot. Stanier 5MT 4-6-0s Nos 45407 and 44871 get away from Sellafield heading the return 1L29 to Carnforth on 5 May 1973. (Author's Collection)

Against a fine Cumbrian Fells backdrop, 46115 *Scots Guardsman* crosses Eskmeals Viaduct heading the return 1Z22 16.29 Ravenglass–Birmingham International 'Lakelander' charter on 19 June 2010.

The unusually late path, a 19.55 departure from Carlisle for Hexham, and a relatively clear sky but with the sun diffused behind a thin strip of perfectly timed cloud, produced an exciting opportunity to capture a *contra-jour* image of the 'Cathedrals Explorer' railtour headed by 60009 *Union of South Africa*. This was the Bangor–Hexham leg of the tour on 13 May 2013, part of an eight-day round Britain railtour originating in the south-east of England.

LMS Jubilee 5690 *Leander* crosses the Eden river bridge at Etterby heading the Glasgow–Liverpool 'West Highlander' charter on 6 September 2010.

Following completion of the 2013 Jacobite season of services based from Fort William, Ian Riley's Black 5 4-6-0s, 44871 and 45407, returned south with empty coaching stock. With open cylinder drain cocks, they get the right away from Carlisle in fading light on 28 October 2013, the 5Z91/5Z92 from Bo'ness to Carnforth and Bury. Despite the modern-day overhead equipment, a pair of Black 5s still looks very much at home at 'The Citadel', a class so synonymous with the Border City.

No. 46115 *Scots Guardsman* crosses the River Esk at Mossband working the 1Z67 Cadder–Oxenholme 'Cathedrals Explorer' charter on 10 May 2011. The concrete bridge, installed at the time of the construction of Kingmoor Marshalling Yard in 1963, with the provision of an additional Up goods line which fed in from the Waverley route at Longtown, is one of two points where the WCML runs only a few feet above sea level.

The Kingmoor shed clock shows 11.25 a.m. and it is also the eleventh hour for the Princess Coronation Pacifics. No. 46226 *Duchess of Norfolk* and 46257 *City of Salford* are facing south ready for their next diagrams during their last year of service. No. 46257 was the last of the class and the only Princess Coronation to be built after nationalisation. They were both withdrawn in October 1964 and disposed of by West of Scotland Shipbreaking at Troon Harbour, 46226 during February 1965 and 46257 during December 1964. (Author's Collection)

In the 1950s the former Caledonian Railway shed at Kingmoor had an allocation of over 140 locomotives and even on the day of closure it had ninety-two locomotives on its books. An evening line-up at Kingmoor on 20 August 1967 sees Jubilee 45593 *Kolhapur*, Black 5 No. 45279, 9F 92110 and Black 5 No. 45021 (far right), one of the first batch of the 842 5MTs introduced in August 1934 and built by Vulcan Foundry , at the north end of the depot. The Jubilee class names commemorated aspects of Britain's great historical past, including the territories of the British Empire. (Author's Collection)

The transitional period from steam to diesel in Cumberland is all too obvious in this summer 1966 scene at Etterby Junction. Brush Type 4s are now handling principal services on the WCML, and diesel traction is maintained alongside steam at Kingmoor depot, although still with much steam traction in evidence. The new diesel depot, now the Direct Rail Services TMD and head office, is under construction on the Down side.

With all track removed, the former steam motive power depot building at Kingmoor awaits demolition in 1970. Very soon the site was to be taken over by birch trees and bracken and now forms a small part of the large Kingmoor Nature Reserve north of the city. (Mike Blenkinsop, Author's Collection)

The Johnson Midland 1F 'half cab' 0-6-0 tank, forerunner of the familiar 'Jinty', was unknown in Cumbria until 2002, the nearest allocation for a member of this class in the 1950s being at Skipton. No. 41708 does not looking out of place on the Derwent River Bridge, Workington, on 15 September 2002, visiting Cumbria for a photographic charter at Corus Workington Steelworks and Cumbria County Council Workington Docks. The diminutive tank has just dropped a consignment of loaded rails from the steelworks for shipment to Ireland and returns a rake of empty internal bolster wagons from the port. The timber trestle bridge was partly washed away by the swollen river during the tragic flooding in November 2009 and has since been demolished. More photos of Workington Steelworks and industrial railways in Cumberland and Westmorland since the early 1960s will appear in a future book by the same author and publisher.

With the pennants bearing the Saltire Cross of St Andrew proudly fluttering in the breeze as if to reaffirm Carlisle's historical connections with Scotland, Stanier 5MT 44871 (with sister 45407 and two support coaches out of view) just catches the last rays of the setting sun as it heads away from the Border City on 29 October 2012, returning from Fort William to Bury (East Lancashire Railway) following completion of the 2012 season of Jacobite services on the West Highland Extension to Mallaig.